Gustavo A. Chaparro-Baquero

Petri Net Workflow Modeling
for Digital Publishing

Gustavo A. Chaparro-Baquero

Petri Net Workflow Modeling for Digital Publishing

Measuring Quantitative Dependability Attributes

VDM Verlag Dr. Müller

Bibliographic information by the German National Library: The German National Library lists this publication at the German National Bibliography; detailed bibliographic information is available on the Internet at http://dnb.d-nb.de.

Copyright © 2007 VDM Verlag Dr. Müller e. K. and licensors
All rights reserved. Saarbrücken 2007
Contact: info@vdm-verlag.de
Cover image: www.photocase.de didi
Publisher: VDM Verlag Dr. Müller e. K., Dudweiler Landstr. 125 a, 66123 Saarbrücken, Germany
Produced by: Lightning Source Inc., La Vergne, Tennessee/USA
          Lightning Source UK Ltd., Milton Keynes, UK

Bibliografische Information der Deutschen Nationalbibliothek: Die Deutsche Nationalbibliothek verzeichnet diese Publikation in der Deutschen Nationalbibliografie; detaillierte bibliografische Daten sind im Internet über http://dnb.d-nb.de abrufbar.

Copyright © 2007 VDM Verlag Dr. Müller e. K. und Lizenzgeber
Alle Rechte vorbehalten. Saarbrücken 2007
Kontakt: info@vdm-verlag.de
Coverbild: www.photocase.de albiedo
Verlag: VDM Verlag Dr. Müller e. K., Dudweiler Landstr. 125 a, 66123 Saarbrücken, Deutschland
Herstellung: Lightning Source Inc., La Vergne, Tennessee/USA
          Lightning Source UK Ltd., Milton Keynes, UK

ISBN: 978-3-8364-1889-8

This work is dedicated to every person that believed in me, specially my Mother Ana Lucía, my Aunt Shirley, my Grandma Aleja, my beloved family and my girl-friend Viky.

ACKNOWLEDGMENTS

Thanks to my advisors Dr. Nayda Santiago, Dr. Wilson Rivera and Dr. Fernando Vega. Thanks to my mother, aunt, grandma, my whole family and my girlfriend, my huge group of friends, PDC and ADMG Labs, HP Labs, the University of Puerto Rico at Mayaguez, Eng. Sameer Handam and thanks to all my sponsors.

This research work has been supported by a grant from the Imaging and Printing Group (IPG) of Hewlett-Packard (HP), Aguadilla, Puerto Rico.

TABLE OF CONTENTS

LIST OF TABLES

LIST OF FIGURES

LIST OF ABBREVIATIONS

DP	Digital Publishing
RIP	Raster Image Processing
WF-net	Workflow-net
PN	Petri Nets
SPN	Stochastic Petri Net.
GSPN	Generalized Stochastic Petri Net.
CTMC	Continuous Time Markov Chain
QDWM	Quantitative Dependability WF-net based Model

LIST OF SYMBOLS

λ_i Exponential distribution rate for each transition firing

P_F Probability of Failure

R Reliability

CHAPTER 1
INTRODUCTION

1.1 Preface

Digital Publishing (DP) is the process of linking printing presses to computers with the purpose of raise the quality level for short-run printing. However, the realization of this potential has been seriously hampered by a number of difficulties. These difficulties include both the problem of getting the document to print correctly without faults and errors on the press and the difficulty of managing the increasingly complex workflow. [1].

The most common format used in graphic arts and printshops is PDF (Portable Document Format), which was not only developed for graphic arts industry, but also for the Internet and enterprise world. For that reason, it is flexible to include or exclude certain objects and parameters (like fonts and sounds), that may be the source of an incorrect job printing. Thus, this flexibility causes the faults and future failures present in the system.

Consequently, digital publishing not only opens up new business but also requires new business models which lead to new workflow designs. The fact that information remains digital from the design stage all the way to printing leads to potential automation of processes that in traditional workshops are still manually executed.

The pre-press process in DP consists on different job treatment stages involving the correct set up of each job in order to be printed. The typical pre-press stages in a DP workflow are described in Table 1–1 [2].

Table 1–1: Digital Publishing pre-press process stages

Stage	Description
Intent	Track the document specifications provided by the client, like type of job (book, brochure, poster, etc.), tolerance of quality (Magazine, Newspaper, flyer, etc.), and due date of the job.
Pre-flight	Check if the digital document has all the elements requited to perform well in the production workflow. These elements include page file format, image resolution, font types, safety margins and mismatched colors.
Trapping	Overlap colors to compensate press registration. Register is the accurate positioning of two or more colors of ink in a printed sheet.
Imposition	Arrange individual pages on a press sheet, so that when it is folded and trimmed, the pages are in the correct orientation and order.
Proofing	Check physically if there are faults remaining in the job verifying the output before printed. Provides an overall view of the color, sizes, and placements of all job elements. Sometimes printshops use proofing printers that simulate the results obtained in the high-volume printers. Conventional way is film-based.
Ripping	Decode Postscript, creates an intermediate list of objects and instructions, and finally converts graphic elements into bitmaps for rendering on an output device. The term *Ripping* is well-known in the DP industry. It is going to be used along this document making reference to the action of RIP (*R*aster *I*mage *P*rocessing) a job document.

There are isolated software tools that process job documents in each stage of the DP workflow for the DP pre-press process. Each of these software packages is designed to complete a stage of the DP workflow, but they do not ensure by themselves the correct printing of a job. There are some packages available aimed at the managing of a DP workflow. However, those packages cannot guarantee the automation of the entire process with an acceptable level of reliability. An industry goal is to integrate the processes into a production workflow or a supply chain path to reduce the costs, increase productivity, and serve customers better [3].

1.2 Problem Statement

One of our goals is to create a workflow process definition of the pre-press DP process, integrating every singular component of each stage into a unique production workflow. The rational is that this workflow definition would improve the global process itself and consequently its dependability. The analysis of dependability is useful if the system or process needs to be critically trustable or if its failures are decreasing the throughput of the system. Besides, to measure dependability quantitatively aids in the analysis of the behavior of the system in the presence of faults. It also estimates which parameters provide the system with a higher trustworthiness. Performance is useful to characterize the system and its throughput, but quantitative measures of dependability shows the probabilistic estimates of the future incidence of the faults. This measures help justify the functional specifications that the system has to meet.

To create a workflow process definition, it is necessary to know the business process definition to be modeled, in order to map it to a workflow model. From the resultant workflow model we should be able to analyze the dependability of the system, measuring quantitatively attributes of dependability such as reliability, maintainability, availability and safety.

1.3 Solution Approach

This work describes the concept of workflow modeling using Petri Nets for the Digital Publishing business process and how the attributes of dependability are measured in a quantitative form, improving the analysis tools necessary to achieve a good workflow process definition. We also propose a methodology for measuring quantitative dependability attributes from a workflow process definition.

1.4 Document Structure

This work is organized as follows. Chapter 2 provides background related to Petri Nets, workflow modeling and dependability. Chapter 3 presents a DP workflow based on the informal description of the business process. Chapter 4 describes the application of an alternative methodology to measure dependability for DP, and the use of Generalized Stochastic Petri Nets for the analysis of workflow model characteristics. Attributes of dependability are measured in a quantitative form, from the workflow model itself to improve modeling. Finally we point out a comparison between results from both methodologies, and concluding remarks in chapters 5, and 6, respectively. Appendix A describes the service manual for the software tool developed for this project that implements QDWMs and appendix B shows the its user manual.

CHAPTER 2
PRELIMINARY CONCEPTS

There are three basic concepts involved in this work: *Petri Nets, modeling of workflows*, and *dependability*. In the following paragraphs these concepts are described.

2.1 Petri Nets

A Petri Net (PN) is a five-tuple (P, T, I, O, MP) where P represents a set of places, $P = p_1, p_2, ..., p_n$, with one place for each circle in the Petri Net graph; T represents a set of transitions, $T = t_1, t_2, ..., t_m$, with one for each bar in the Petri Net graph; I represents an input function that defines directed arcs from places to transitions; O represents an output function that defines directed arcs from transitions to places; and finally MP represents the marking of places with tokens. Tokens are represented as small dots or integer numbers and the diminution of tokens over the places determine the state of a Petri Net. An example of a Petri Net is shown in figure 2–1, where the circles represents places, the boxes represents transitions and the tokens are represented by small dots.

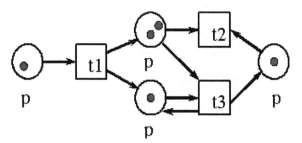

Figure 2–1: Example of a Petri Net (PN)

Transitions are the active components of a Petri Net. A transition could represent: events, operations, transformation, transportation and so on. Places are passive and they could represent: medium, buffer, geographical locations, phase, and conditions. Tokens often indicate objects (physical or representing information). The dynamic behavior of the system is then modeled by the flow of token and firing of transitions.

As a graphical tool, Petri nets can be used as a visual-communication aid similar to flow charts, block diagrams, and networks. In addition, tokens are used in these nets to simulate the dynamic and concurrent activities of systems. As a mathematical tool, it is possible to set up state equations, algebraic equations, and other mathematical models governing the behavior of systems.

Petri Nets provide a uniform environment for modeling, formal analysis, and design of discrete event systems. Petri Nets models are used for the analysis of behavioral properties and performance evaluation, as well as for systematic construction of discrete-event simulators [4]. In a discrete event system, discrete entities change state as events occur and the state of the system changes only when those events occur. Digital Printing involves a combination of separate stages that manipulates a job in order to ensure its printability and its correct delivery to the client. The arrival, manipulation and print out of those jobs are discrete events [2]. For that reason Digital Publishing is modeled as a discrete event system.

Petri nets have evolved to incorporate more detailed techniques for modeling. Those techniques have been called *Extensions*.

2.1.1   Petri Net Extensions

Since their invention in the early 60's, Petri Nets theory have been increased with new ideas, which means that a Petri Net could be enhanced with different type of extensions. A *High level Petri net* is a net that involves three extensions useful for describing workflow models and mapping business process to them. The

6

three variants are: Colored Petri Nets, Hierarchical Petri Nets, and Petri Nets with time. *Colored Petri Nets* associate a different color to each token generated for each case routed through the net. Thus, a token color is associated with singular characteristics depending on the case. *Hierarchical Petri Nets* associate subnets to some transitions of a main net. This helps build and organize easily a net because each subprocess of a main process could be represented with a subnet of a main net. The final variant, *Petri nets with time*, have had many approaches. One of them associates a variable of time with each token. This number acts as the firing enabling time of that token, which means that, for instance, a token with enabling time of six cannot be fired before six units of time after its arrival to the place. This is helpful to model the timing of a process [5]. Another approach for a Petri Net with time, *Timed Petri Nets*, associates time with a firing delay on each transition in the net.

2.1.2 Timed Petri Nets

To study performance and dependability issues of systems it is necessary to include a timing concept into the model, because an ordinary PN only describes the structure of the model, but performance and dependability analysis involves also time evolution study. There are several possibilities to do this for a Petri net. However, the most common way is to associate a firing delay with each transition. This delay specifies the time that the transition has to be enabled, before it can actually fire. If the delay follows a random distribution function, the resulting net class is called stochastic Petri net. Different types of transitions can be distinguished depending on their associated delay. These include immediate transitions (no delay), exponential transitions (delay is an exponential distribution), and deterministic transitions (delay is fixed). An example of a Timed Petri Net is shown in figure 2–2.

As it was mentioned before, the dynamic behavior of the system is determined by the movement of tokens through the net. This movement is product of transition

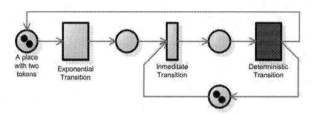

Figure 2–2: Example of a Timed Petri Net

fire enabling. A transition is enabled to fire when every input place associated with it has the number of tokens that each arc specifies. When any transition fires tokens, those that were at the input places are removed, but just one transition is enabled to fire at a time. The multiplicity of each arc, which is the number of tokens to be removed or fired, is specified as an integer close to the arc, but if it is not specified, its default value is one.

Each configuration or distribution of tokens along the Petri Net represents a state of the net, and this state is called a *marking*, thus each transition firing generates a marking of the net. If the number of tokens is finite or bounded, then the number of markings is too. If a marking M_x could be obtained from an initial marking M_i through token firings, it is said that the marking M_x is reachable from the marking M_i. The reachability set is a graph where the initial node represents the initial marking of a net, M_i and each subsequent child of this node represents the reachability marking after firing each transition from a specific marking. An inhibitor arc is an arc that will cause a transition firing, if the number of tokens in the input place indicated by its multiplicity are not in this input place. If the inhibitor arc is connected to an output place, this arc will remove the tokens fired from the net, avoiding them to reach the output place. In figure 2–3 an example of a Petri Net and its corresponding Reachability Graph is shown.

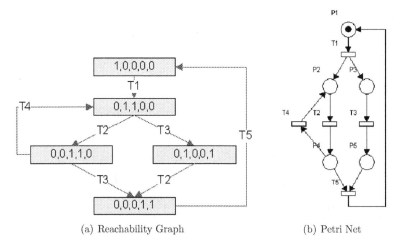

(a) Reachability Graph   (b) Petri Net

Figure 2–3: Example of a Reachability Graph

Stochastic Petri Nets

A *Stochastic Petri Net* (SPN) has associated a firing delay to all of its transition, and this delay of time is associated with a random variable exponentially distributed. This means that the distribution of the random variable X_i of the firing time of a transition is given by $F_{X_i}(X) = 1 - e^{-\lambda_i \cdot X}$. The average time of firing of the transition t_i is $\frac{1}{\lambda_i}$. The qualitative analysis of a SPN is made analyzing the Markovian process associated with the SPN itself. This is done by adding to each arc of the reachability graph, a weight equivalent to the exponential distribution rate (λ_i) of each transition firing. This results in obtaining a Markov chain from the SPN. Achieving the steady state distribution of the Markov chain, is possible to compute performance measures like the probability of being in a subset of markings, the mean number of tokens and the probability of firing any transition.

Generalized Stochastic Petri Nets

Stochastic Petri Nets are helpful for evaluating in terms of probabilities the extent to which some attributes like availability, maintainability, safety and reliability are satisfied into a system [6]. It is not always useful to associate a random

9

distribution function of time to each firing transition in the net, because either the execution time of this transition is zero (immediate) or this execution time could be approximated to zero. The inclusion of immediate transitions makes it easier the analysis of the net reducing the states that have to be computed. A Petri Net that involves exponentially distributed transitions and immediate transitions is called a *Generalized Stochastic Petri Net* (GSPN) [7]. Figure 2–4 shows the process of obtaining a Continuous Time Markov Chain (CTMC) from a GSPN. First, the corresponding Reachability Graph is obtained from the GSPN. Then this reachability graph is converted into a CTMC.

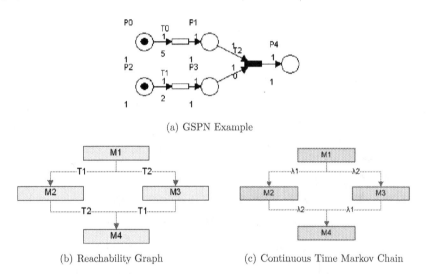

(a) GSPN Example

(b) Reachability Graph   (c) Continuous Time Markov Chain

Figure 2–4: Example of how to obtain a Continuous Time Markov Chain from a Reachability Graph of a GSPN

2.2   Workflow Modeling

Workflow is referred to the study of operational aspects of a specific activity in a workable environment. This includes how to arrange tasks, how to perform them, how to establish their order of execution and how to supervise them. There are different concepts that must be kept in mind in order to model a workflow. A *case* is defined as the tangible object that is processed or modified in the workflow. Each case has a beginning into the workflow and therefore and end, and each case can also be differentiated from another case. A *task* is a part of the whole work that has to be done in order to process or modify each case. A task is executed by a resource, which means a person or a machine and it could be seen also as a process that cannot be subdivided. A *resource* is referred to every person, machine or groups of these, that perform a specific part of the work, a task. Each case incoming to the workflow involves a sequence of tasks to be performed in order to treat the case itself. This sequence along with the conditions that determine the order of execution of these tasks is called a process. *Routing* refers to the path that each case takes into the workflow, which is associated with the order of the tasks that treat that case [5].

The concept of workflow management refers to the action of taking the business processes out of the applications and put them in decomposed management systems. A workflow system manages the workflows and defines the routing of the information of each case through the human resources and the application programs. A Workflow management system (WFMS) is used to define and use workflow systems. In general a workflow management system controls the workflows that involve each case management as well as the resources and applications.

Frequently PN are used to model workflows, since it supports modeling of the dynamic changes of a workflow system. Within a workflow it is necessary to handle cases, which are individual activations of a workflow. Each case fires different tasks in different order and each task has preconditions that must be accomplished before complete it.

There are at least three good reasons to use PN in workflow modeling and analysis. First, PN use formal semantics despite their graphical nature, which guarantee a structured definition of the model. Second, PN are state-based instead of event-based, which allows to model the state of each case clearly. Finally, there exist abundance of analysis techniques for PN. Among them, we can count analytical tools and simulators [8]. Although there are many different ways to define a workflow process, their expressive power is often weaker than the expressive power of PN, because PN are able to model certain routing structures that other models cannot model.

Many classes of PN for workflow modeling have been proposed. One of those classes is the Workflow-net, which is an extension of a PN proposed by Wil van der Aalst [5].

2.2.1 Workflow-nets

In a Workflow-net must exist a place with no incoming arcs, which identifies the beginning of the process, and a place with no outgoing arcs which identifies the end of the process. A workflow-net must be strongly connected, which means that any node can be reach from the starting place following a certain path.

The theory of Workflow-nets has additional classes of transitions that aids to clarify the routing rules described by the workflow model. These transitions are AND-split, AND-join, OR-split and OR-join, and they are shown in figure 2–5 along with their corresponding PN meaning. In table 2–1 each of these kind of transitions is described.

Table 2–1: Description of the additional transitions of a Workflow-net

Name	Description
AND-Split	A token must be produced for each of the output places under all circumstances.
AND-join	The task can only take place once there is a token at each of the input places.
OR-split	A token must be produced for just one of the output places. A decision rule must be adopted to solve the corresponding firing conflict.
OR-join	The task take place once the single token reach one of the input places.

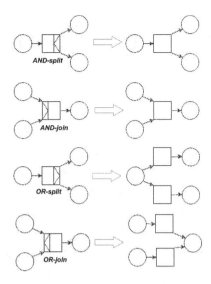

Figure 2–5: Additional transitions of a Workflow-net and their corresponding PN meaning

A Workflow-net must be *sound*, which means that it must not have unnecessary tasks and every case treated by the process must not make any reference to this case once the case reach the final state, i.e., remaining tokens must not be leaved in the process. Thus, the workflow-nets theory define that a net is sound if it fulfills the following three requirements:

- For each token put in the place *start*, one (and only one) token eventually appears int he place *end*
- When the token appears in the place *end*, all the other places are empty
- For each transition (task), it is possible to move from the initial state to a state in which that transition is enabled

The soundness property corresponds with two additional properties of a PN: *liveness* and *boundedness*. PN theory identifies certain properties that a net could have. One of those conditions is to be *Live*. A PN is live when it is possible to reach (for each transtition t and for every state reachable from the initial one) a state in which transitionn t is enabled. Another condition is to be *Bounded*. A PN is bounded when there is an upper limit to the number of tokens in each place. When a PN is bounded to just one token per place, the net is also safe. Thus, a PN is *Safe* when the number of tokens in each place will never be larger than one. Thus, if we have two sound and safe Workflow-nets V and W and we have a task t in V which has precisely one input and one output place, then we may replace task t in V by W and then the resulting Workflow-net, X, is sound and safe again, and therefore X is also sound. The crux of the proof is the observation that every state in the resulting Workflow-net X can be mapped onto a state in V and a state in W and vice versa. For more details we refer to [5].

In order to build a sound Workflow-net, its construction must be done with sound processes. The theory identifies four basic constructions for routing tasks, which fulfill with the soundness property. These constructions are described in figure 2–6. These four *buildingblocks* could be seen as the algebra for Workflow-nets construction, which means that all the properties of the *buildingblocks* are inherited by every Workflow-net resulting from constructing with them. It is noteworthy to say that a Workflow-net could be made also using any sound and safe Workflow-net.

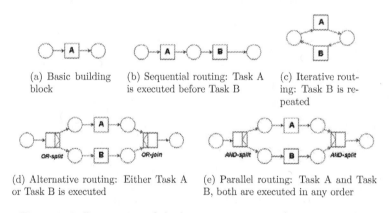

(a) Basic building block

(b) Sequential routing: Task A is executed before Task B

(c) Iterative routing: Task B is repeated

(d) Alternative routing: Either Task A or Task B is executed

(e) Parallel routing: Task A and Task B, both are executed in any order

Figure 2–6: Description of the basic constructions for routing tasks

2.2.2 Measures Taken from a workflow model based on Petri Nets

Workflow modeling and analysis based on PN have been used in many settings in industry. In many opportunities, PN analysis has helped to verify the soundness of the model and repair errors in the PN itself, proving that the implementation of PN in the modeling and analysis of workflow systems provides a standard design method approach [9]. Some of those studies have been concentrated in workflow performance issues [10].

A PN which models the control-flow dimension of a workflow is called a *Workflow-net*. A Workflow-net specifies the dynamic behavior of a single case in isolation. In [11] an approach is presented to extract both a workflow model and performance indicators from timed workflow logs, which means to track the completion of a task in time and to track the route of each case through the workflow.

Not only the task and resources are important to be managed in a workflow process, but also the time of completion of those tasks is important. Time management is essential in determining and controlling the life cycle of each activity involved in the business process. The so called Workflow-nets (WF-nets) have been extended with time intervals in order to make them able to analyze workflow systems with time constraints. Those time intervals in this Timed Workflow-nets (TWF-net) are associated with the transition firing, but unlike the stochastic PN, which use a random variable, they use a deterministic interval [12].

Workflow models based on PN have been used to design workflow process definitions of various business processes with successful results. Because the workflow model is the heart of the workflow management system, it must be a careful design work. PN theory have been used only to debug the model itself and to analyze it, which involves adding time variables to the net, and measuring performance issues. It has been used to improve the model in early stages of the model creation. However, the analysis made over a WF-net based on PN reduces to soundness and performance study of the model, but other measures like dependability has not been introduced in that examination of the workflow model.

2.3 Dependability

Dependability is the ability to deliver service that can justifiably be trusted. This concept includes measures such as reliability, availability, maintainability, and safety. The methodology behind dependability focuses on managing errors, which means to identify, treat and classify the different types of errors that could be found in a system. This concept formulates a series of definitions of errors and failures that exist in the system and, in general, dependability is a sum of attributes for error identification of a system.

The definition of the dependability concept matches with what we want the digital printing workflow to be. The metric establishes that errors are acceptable, in a certain level for the system, but the system itself must ensure that those errors will not cause any problem to the user. Only in that case, the user will trust the system. Certain levels of dependability are necessary to guarantee in a DP workflow process. Besides, the digital printing system is highly based, on the warranty that could be offered to the client. This means that the final job will be finished on time and correctly and with the least amount of acceptable errors. The system is based on the trustworthiness that each client gives to the system.

There are three concepts that shape dependability, which are the attributes of, the means to attain and the threats to dependability. The first aspect of dependability model that we have to point out is the way that it treats and classifies the possible errors and faults present in any system. Thus, the threats to dependability are shown as faults, errors and failures. The next aspect refers to the attributes or concepts that integrate dependability, which are basically availability, reliability, safety, confidentiality, maintainability and integrity. Depending on the system which is going to be adapted to dependability, the above attributes will be included in the measures or not. The last concept refers to the means to attain dependability which consist of fault treatments, either at the beginning of the design or in the analysis of the future behavior of the system. The means to attain dependability are: fault prevention, fault tolerance, fault removal and fault forecasting. Figure 2–7 shows the taxonomy of dependability. A very important thing is the fact that the whole model to attain dependability of a system is a framework, which needs the addition or subtraction of certain necessary aspects depending on the particular system that is required to be designed, but there are a minimum of aspects that must be included.

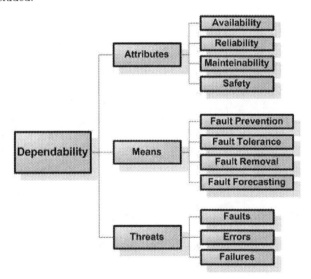

Figure 2–7: Dependability Taxonomy

As mentioned above, there are four means to attain dependability in a system, each of them used in a different stage of the establishment or design of the system. Those means are: Fault Prevention, Fault Removal, Fault Tolerance, and Fault Forecasting. *Fault Prevention* refers to the avoidance of faults in the fists stages of the system, which means system entries with less or no faults. *Fault removal* refers to verify the system looking for faults and correct them. *Fault tolerance* refers to make the system strong enough to detect a fault or an error, and recover from it by itself. *Fault forecasting* refers to performing an evaluation to the system behavior with respect to fault occurrence or activation. Evaluation has two aspects: To identify and classify possible faults and errors that the system could show is a qualitative evaluation. To establish in terms of probabilities the extent to which the attributes pertinent to the evaluated system are satisfied in the presence of those faults and errors is a quantitative evaluation.

Fault Prevention and Fault Removal for the DP pre-press process are not meaningful to be studied, because it is assumed that each job incoming to the press will contain faults and errors making the reduction of the severity of those faults and errors difficult. For the DP pre-press process could be possible to study a fault tolerance mean to attain dependability, perhaps using common techniques such as adding replication or enhancing the system to do exception handling [13], but that study is out of the scope of this work. This work is focused on obtain qualitative measures of the attributes of dependability from a workflow process definition, particularly from the workflow process definition of the digital printing pre-press business process.

There are known tools and techniques for quantitative dependability analysis such as: Static and dynamic fault trees, Stochastic Petri Nets, Markov and queuing models, and Reliability block diagrams [14]. Fault Tree Analysis (FTA) and Reliability Block Diagrams (RBD) are *Combinatorial Models*. Combinatorial Models enumerate all the possible combination of failed and working elements or events that represents either the success of the failure of the system. Petri nets and Markov models are *Noncombinatorial Models* [15].

Unlike Markov models, combinatorial models cannot accurately model dynamic system behavior, making this Markov models the most common way to study the dependability of complex systems [16]. Besides, markovian models have more powerful approaches than combinatorial models, but are more complex too. They can model behavior of the system that cannot be modeled with combinatorial models such as time [17].

A Fault Tree Analysis is intended to model the combination of conditions that result in the system failure and is a well known and a common technique to do a dependability analysis. A fault tree diagram uses "AND" gates (all inputs must fail for the gate to fail), "OR" gates (any input failure causes the gate to fail), and "K-OUT-of-N" gates (k or more input failures cause the gate to fail). Every input to the tree is known as a basic event and there is a single output called the top event representing a system failure event. Figure 2–8 shows an example of a fault tree.

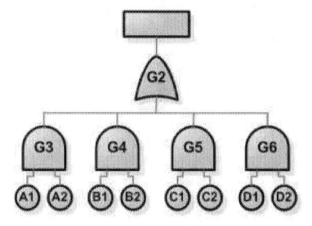

Figure 2–8: Fault Tree Diagram example

Another common way to do a dependability analysis is to do it over a *dependability model*, based on a Markov model in which each node is associated with the failure of a component in the system and the weight in the arcs are associated with the failure rate parameter. Furthermore, this Markov model could be associated with an Stochastic PN or viceversa. Over this dependability model different kinds of studies could be done, for instance, a preliminary evaluation, a sensitivity analysis that explores the effects of changing the model, and a specification determination

19

that establish the initial conditions to achieve a known result [18]. Stochastic models have been widely used in engineering to describe the operation of a system with respect of time in order to analyze its reliability, maintainability or safety, because these Markov methods have shown to be powerful tools in this kind of analysis.

Different methodologies and strategies to develop and construct dependability models have been proposed, based on stochastic PN or stochastic reward nets, as in [19], [20] and [21]. Some of those strategies have become into automatic analysis software tools. For instance, one of the applications of the project HIDE [22] transforms UML to SPN in order to create a dependability model to analyze the reliability of the system with the intention of improve the system design. MEADEP [23] provides some additional features over the dependability analysis. Another example of dependability modeling tool is web-based tool aimed to model dependability based on a framework supported on the values of acceptable manifestation of failures that are gathered from the user [24].

In [25] a modeling methodology is presented, in which a design-engineer can specify a design using a single model which is capable of combine both the performance and dependability characteristics of a system. The main feature of this methodology is that a single model is used to study performance, reliability and behavior, eliminating inconsistencies between different models. The above mentioned methodology builds an analytical model using Colored Petri Nets from a block model, which is used to analyze behavior, performance and dependability. Such a methodology is proposed for the design of hardware devices using also hardware description languages. ADEPT [26] is a tool developed to implement a digital systems design environment, which incorporates the methodology described in [25].

CHAPTER 3
DEVELOPING A WF-NET MODEL FOR DIGITAL PUBLISHING

From the informal description of the business process of DP, it is possible to identify each part of the workflow model such as tasks, cases, processes, and routes, and then, construct the workflow model. Thus, this model can be introduced into a workflow engine of a workflow management system. The following paragraphs depict the essential stages in a Digital Printing process [2].

When a job comes into the printshop, this job has a minimum of specifications. Some of these specifications are the type of job (book, brochure, poster, etc), the quality of service that the client needs for his/her job (Magazine, Newspaper, flyer, etc) and the due date of the job. This information must be collected from the client and well wrote down in order to track the printing process of the job. This first stage of the process is what we call the *Intent* stage.

After receiving a job, the printshop must ensure that it does not have faults, and the printshop does it processing the job in the *Preflight* stage. This stage is aimed to check a series of minimum items that the job must contain in order to be well printed, depending on the specifications before registered in the previous stage that received the job from the client. The preflight stage selects the best profile for the job treatment. A profile is a set of characteristics that will be checked in the job, for example, black and white, generic press, CMYK color printing, and so on. A report of the process is made and this report indicates which and where the faults in the job are, so they can be fixed either in the preflight station or in other station better qualified. The preflight technician decides based on the report whether a fault is fixable or not. Sometimes, there are catastrophic faults present in the job, which means that it will never be well printed or could be printed with less quality that the specified for the job. Some of these faults become errors, because they cannot be fixed. If a fault or an error cannot be fixed, the job must be submitted back to the client. The client will then repair the job or will send the object that contains the problem in the job, so the printshop could be able to do it. For instance, if a magazine publication printed in color has an image in one of its pages with 300 dpi (or less), it could be considered as a fault that has a highly probability of become into an error, because the quality of the color printing of the magazine should be high and that image will probably show a pixelation effect (a pixelation effect is when the eye can notice the color change in the dots of the image easily).

Once the preflight stage checks and fixes all the faults in the job, it is submitted to be trapped. There are a variety of software packages that are made to trap a job. The *Trapping* stage consists into fill the border of a color-change line with some of the same color tone. Due to the high production volumes and the speeds of the printing machines, sometimes the paper could change its position during the printing itself, so, for instance, a job containing a black object so close to a yellow object, and this job is not trapped, it could be printed with a gab between the black and the yellow objects; unless the color of the sheet of paper is yellow, this gap will get into the notice of the reader of this job. If the job of the above example is trapped, even if the paper moves, the gap will be filled with the same color tone of one of the objects. The trapping process could be applied to more than two colors at a same time.

In order to check the proof of accuracy of the job that will be printed, it is sent to the preliminary *Proofing* stage. In this stage a printed copy of job is made to check physically if there are faults remaining in the job. This proof also provides an overall view of the color, sizes, and placements of all job elements. Sometimes printshops use proofing printers that simulate the results obtained in the high-volume printers. If an error is detected the job is sent back to the preflight stage in order to correct it, and again, an expert or a technician is responsible to judge such a decision, based on the physical printed copy.

The following step is to impose the job. *Imposition* is the stage where each page of the job is arrange in a determined position into the whole big sheet of paper that will be sent to the high-volume printer. In the imposition stage, called the finishing stage, are also set the margins and conditions of the job in order to be well folded and trimmed. There is a variety of types of folds that a document could have, for instance, the French fold, work and turn, sheet wise or work and flop. The document is also imposed depending on the final binding, for example, the job could be bind using wire stitches, cloth tape, spiral plastic coil or a perfect binding. . There are other types of proofers, which involve calibrated monitors instead of printed copies, this is called soft proofing. A final proof of accuracy is made after the imposition stage. This proof is a legally binding sample of how the job is expected to appear when printed.

Finally, when the job is supposed to be completely free of faults and errors, it is sent to the *RIP* or *Ripping* stage. RIP is the acronym for Raster Image Processing and it is defined as the action of turn vector digital information (a postscript of a PDF file, for instance) into a high-resolution raster image, which means that RIP takes the digital information about fonts and graphics that describes the appearance of the file and translates it into an image composed of individual dots that the printing device can print. The RIP process can take a long period of time, and if the files of a job sent to be ripped contain faults or errors, the RIP process could fail, making a wrong translation to dots printable by a printer, which means, for instance, fonts overlapped, images with bad resolution or objects misplaced. If a job fails ripping, some procedures could be applied in order to fix the problem. Nevertheless, if the RIP process definitely fails, the job should be sent to an early stage in order to determine the exact problem cause and correct it. Besides, a document job could be labeled as a failure job and it must be sent back to the client.

Based on this description of the system it is possible to map it into a workflow model based on Petri Nets. Dependening upon the informal description of the system, we have created the corresponding process, mapping this description of the system into a workflow model based on Petri Nets. The result is shown in figure 3–1. The boxes marked with a rectangle inside indicates that this task is a subprocess. In the following paragraphs we explain some subprocess corresponding to some stages.

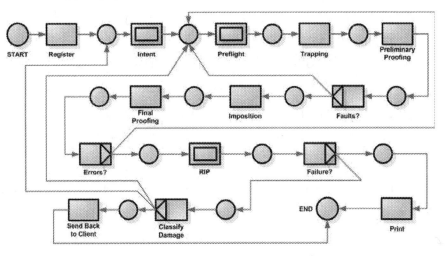

Figure 3–1: Workflow model for the Digital Publishing pre-press process

In figure 3–2, we can see the model representation for the intent sub-process. When a job arrives to the print shop it has to be checked and labeled. The files that make a job can be received by different ways, magnetic disk, optical disk, e-mail, and so on. The intent sub-process looks for to carry out all these tasks. Here the fonts used in the documents are checked to see if they exist in the house, otherwise the client must be informed of that situation. The graphics must be correctly linked in the documents. If the files are compressed, they must be uncompressed to be processed. If the whole job is complete, the document is registered and labeled. In this procedure, the job is classified depending on its type. Besides, the job is rated with a severity grade that will serve as a measure of its final quality requirements. Finally, a due date is assigned to the job.

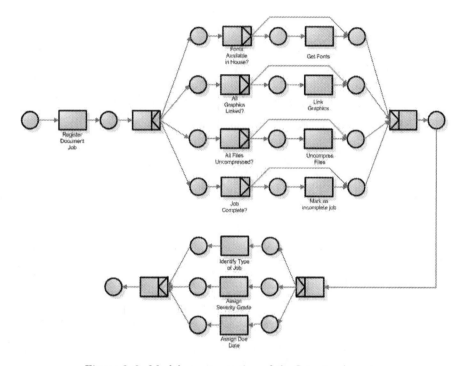

Figure 3–2: Model representation of the Intent subprocess

The first stage that we modeled serving as our first case of study was the pre-flight pre-press process. Figure 3–3 shows the model representation for the preflight sub-process. The job is registered in the preflight module, then a preflight report about it is made, according to a preflight profile preselected. The report is reviewed by a preflight technician or printing expert. The technician search and repair, if possible, faults in the document. For instance, faults related with fonts, image resolution, or color bases. In subsequent chapters we discuss in detail the preflight stage.

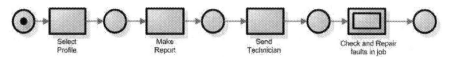

Figure 3–3: Model representation of the Preflight subprocess

The imposition subprocess model is indicated in figure 3–4. It can be seen that each job has to be imposed using one of four possible methods: French fold, work and turn, sheet wise and work and flop. After this procedure, the binding process must be selected for the job, choosing from four options: wire stitches, cloth tape, spiral plastic coil and perfect binding.

The RIP process is shown in figure 3–5. Here the job is received at the stage and is RIPed. Subsequently, it is verified if the RIP process was successful. If it was, the job is sent to be printed at the press, otherwise the job is checked to find errors that cause the RIP to fail. They can be easy or difficult to track down and correct. There are some steps that can be done in order to identify these errors. Here we mention only four of them. Find and replace the font that is causing the problem in order to replace the file of that font with a new and uncorrupted copy. Try to reduce the amount of pages that are printed at a time, dividing the job into smaller ripping packages. Check that all graphics and complex image effects are converted to a bitmap format before placing them into the application file. Finally, recreate the document in a different application, distiller for instance. If an error is identified, but it cannot be repaired, the job must be marked as failure. Besides, if no errors are found to be the cause of the ripping failure, the job must be marked as failure too.

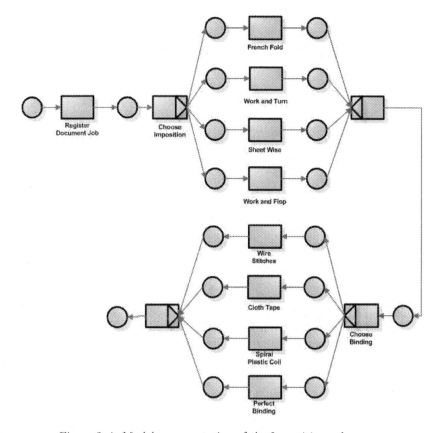

Figure 3–4: Model representation of the Imposition subprocess

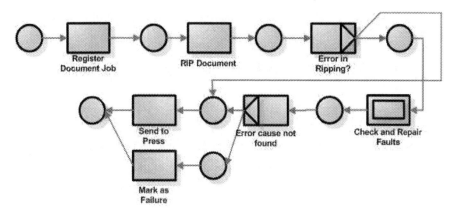

Figure 3–5: Model representation of the Ripping subprocess

We verified the properties of our DP Workflow model and subprocesses using a couple of software tools: WoPeD 1.0 [27] and Woflan 2.2 [28]. WoPeD was developed at the University of Cooperative Education (Berufsakademie) Karlsruhe. It is a tool for editing and simulating WF-nets, that uses Woflan as analysis tool. Woflan was developed at the Technische Universiteit Eindhoven. It checks for soundness in WF-nets. A screen-shot showing the DP Workflow-net of figure 3–1 modeled into WoPeD, is depicted in figure 3–6.

Woflan can verify the soundness of a given process definition. This soundness property is the minimal requirement any workflow process definition should satisfy. The diagnosis view of the software tool (figure 3–7) shows all properties of the process definition in a tree-like manner. At the root, the name of the process definition file is shown. This root node has two child nodes: the upper for the diagnosis results and the lower for the diagnostic properties. The diagnosis results node shows in brief the results on the main properties (workflow, safeness, liveness, soundness). Figure 3–7 depicts a screen-shot showing the soundness analysis done over the DP Workflow-net by Woflan. It can be seen that the Workflow-net is completely sound, because every property and condition for a proper workflow is corroborated to appear in the workflow analyzed.

PN theory identifies certain properties that a net could have in order to fulfill certain conditions. One of those condition is to be *Live*. A PN is live when it is possible to reach (for each transtition t and for every state reachable from the initial one) a state in which transitionn t is enabled. Another condition is to be *Bounded*. A PN is bounded when there is an upper limit to the number of tokens in each place. When a PN is bounded to just one token per place, the net is also safe. Thus, a PN is *Safe* when the number of tokens in each place will never be larger than one.

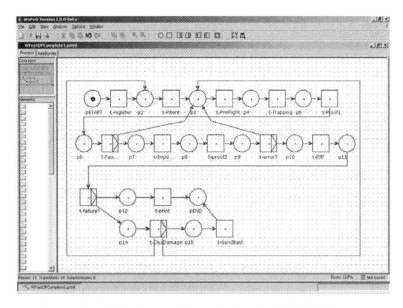

Figure 3–6: WoPeD screen-shot showing the DP Workflow-net

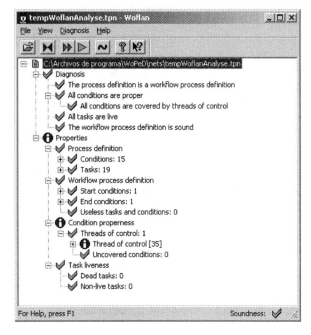

Figure 3–7: Woflan screen-shot showing the analysis done over the DP Workflow-net

CHAPTER 4
MEASURING DEPENDABILITY
ATTRIBUTES USING A WORKFLOW MODEL

We obtained values of dependability attributes in a quantitative form from the workflow model instead from a combinatorial model. In order to do this, we adapted some methodologies and strategies that build models to the creation of the workflow process definition. In order to obtain values for dependability attributes in a quantitative form, from a workflow model instead of from a combinatorial model model, we adapted some methodologies and strategies that build models that allow to measure dependability, to the creation of the workflow process definition.

This approach will lead to the addition of parameters to the model, in order to make it more accurate. Constrasting the analysis provided in [10] and [4], our work proposes an analysis of quantitative dependability over a workflow model, and not only an analysis of its structure and performance. Furthermore, we suggest a methodology to include parameters for dependability analysis into the workflow process definition mapped from the business process of the high dependable system. We want to obtain the qualitative dependability attributes of the system making a *fault forecasting analysis*, which is performed by evaluating the system behavior with respect to fault occurrence or activation. The qualitative evaluation of a system aims to identify, classify and rank the failure modes (the different ways a system can fail) or the event combinations (component failures or environmental conditions) that would lead to system failures.

In the following sections, we present our methodology applied to two subprocess of the DP Workflow-net: The pre-press processes of Preflight and Ripping.

4.1 The *Preflight* subprocess

An study over a variety of PDF documents produced by a diversity of software tools was done in order to identify the most common critical faults present in a job. For this, the reports produced by a commercial preflight tool were analyzed, finding that faults related with not embedded fonts, low image resolution, objects overlapping safety zones and images using the wrong color base were the most common ones. According to our study, the mean probability of find a fault related with fonts not embedded is 67%, the mean probability of find a fault related with a wrong color base is 38%, and the mean probability of find a fault related with a low image resolution is 61%.

Table 4–1 shows the most relevant faults that could be present in a document (based on [2] and on our own study). Faults are classified by the kind of failure that they are able to generate. For instance, a fault related with fonts will lead to a failure that could be classified in domain as a contain failure (C) or a timing failure (T). A contain failure refers to a failure in the content of the service (in DP, for instance, if an image is printed out of the paper margins) and a timing failure refers to a failure in the time of completion of the service (in DP, a job that takes an overdue time in being completed).

The consistency (C) or inconsistency (I) of a failure is seen in if this failure is perceived by all final users in the same way or not, respectively (in DP, a change in the font type could be seen in a different way by each client, and it depends on the clients job, for example, a brochure, a book or a magazine). The consequences of a fault could be classified in minor, medium, and catastrophic, depending on the severity of degradation in the final service provided. Faults such as incomplete or corrupted files could lead to catastrophic failures (C). In contrast, if an image placed

31

in a document requires being in CMYK color process, and is in RGB, these fault could lead to a medium (Md) or minor (Mn) failures.

Table 4–1: Analysis of the possible faults present in preflight

	Domain	Consistency	Consequences
Not embedded Fonts	Content/Timing	Inconsist.	Medium
Low image resolution	Content/Timing	Inconsist.	Medium
Wrong color base	Content/Timing	Consist.	Minor
Missing images	Timing	Consist.	Catastrophic
Incomplete or corrupted files	Timing	Consist.	Catastrophic

In a previous chapter we showed the preflight subprocess, arguing that it was going to serve as our first case of study. We first show an alternative model to measure quantitative dependability attributes from the workflow model of this process. Then, we illustrate our methodology to create a workflow model able to measure quantitative dependability attributes from this system.

4.1.1 Alternative model to measure dependability applied to the Preflight process

We have studied dependability analysis over the workflow model of DP applying either a dependability model as well as a FTA technique. In a workflow model the time associated with each transition models the time consumed by the execution of that specific task. In contrast, the time that a dependability model assign to each transition represents the time that a fault or error takes to be present in the model. A workflow model treats cases and the faults and errors must be associated with those cases. Thus, those faults and errors are inherent to each case an do not depend of time. For a DP case, i.e., a document job the faults and errors have an inherent probability of occurrence independent of time. For this reason, we conclude that a dependability model is not useful to be applied in order to analyze the dependability of the workflow model, because such a model obtained in terms of GSPN would result into a model with a vanishing initial marking. Therefore, we use the FTA technique in order to aid the construction of our model and its corresponding methodology, as well as to contrast the results obtained from it.

In this case of study we assume that the pre-press process of preflight has five fault sources that could lead to a failure of the system, i.e., the pre-flight pre-press process rejects the job analyzed. Figure 4–1 shows the fault tree representation of the preflight subprocess. The couple of events A, B and C represent faults that are fixable in the preflight station. A represents faults related with fonts, B faults related with image resolutions and C faults related with wrong color bases. The first event of each couple represents the presence of the fault in the job and the second event of the couple represents the ability of repair that kind of fault. Events D and E represent non-fixable faults. D represents faults related with corrupted files and E faults related with missing images.

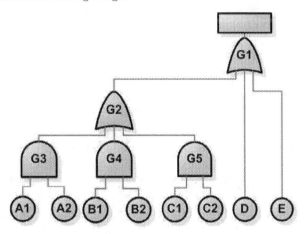

Figure 4–1: Fault Tree for the Preflight pre-press process

The analytic equation that describes the probability of failure of the preflight pre-press process, analyzing the fault tree in figure 4–1, was obtained. Thus, the expression of the probability of failure of the preflight stage is the equation (4.1). We used the corresponding letter instead of use the events multiplication (for instance: $A \Rightarrow A_1 \cdot A_2$)

$$P_F = (A + B + C - AB - AC - BC + ABC) \cdot$$
$$\cdot (1 - D - E + DE) + D + E - DE$$

(4.1)

The equation for the reliability of the preflight stage is: $R = 1 - P_F$. Replacing into this equation, we have:

$$R = 1 - [(A + B + C - AB - AC - BC +$$
$$+ ABC) \cdot (1 - D - E + DE) + D + E - DE]$$

(4.2)

4.1.2 Develop a WF-net including fault parameters for Preflight

To analyze quantitatively dependability attributes in a WF-net, it is necessary to add the identified possible faults of the system into the model generated. We include those faults into the model of figure 3–3 by replacing the task *check and repair faults in the job* by five parallel sub-processes, using an AND-split/AND-join construction. Each of them check the presence of a kind of one of the most relevant faults that a job could have. The resulting net is shown in figure 4–2

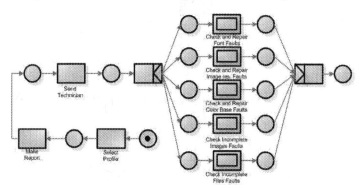

Figure 4–2: Replacing the subprocess "check and repair faults in the job" in the preflight stage WF-net (figure 3–3)

We defined a fault treatment sub-process, shown in figure 4–3. For constructing this WF-net we add an OR-split that checks for the presence of faults. Later, we include another OR-split checking if the fault can be repairable. Finally, we put a task for each of the three situations (no faults, repairable and not-repairable). Subsequently, we replace each sub-process in figure 4–2 for our fault treatment sub-process. This give us as result, a better WF-net for the preflight prepress process, which is shown in figure 4–4

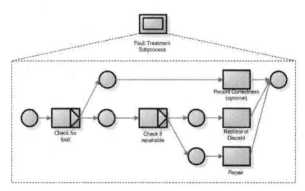

Figure 4–3: Fault treatment subprocess

In the same manner that we test correctness in the DP workflow, we test the correctness of the WF-net for the preflight pre-press process using WoPeD and Woflan. The WF-net for preflight complies with workflow model parameters.

To obtain either performance or dependability measures from a WF-net model, time and fault parameters must be included. Our methodology includes time as a random variable exponentially distributed on some transitions of the net, and include fault parameters as immediate transition weighs. Consequently, the net must be converted into a GSPN. To do so, it is necessary to replace the OR-split blocks (figure 4–5.a) by a combination of transitions. The outcome is governed by a fault probability . The OR-split block (figure 4–5.a) is replaced by an exponentially transition simulating the action followed by two immediate transitions disposed in

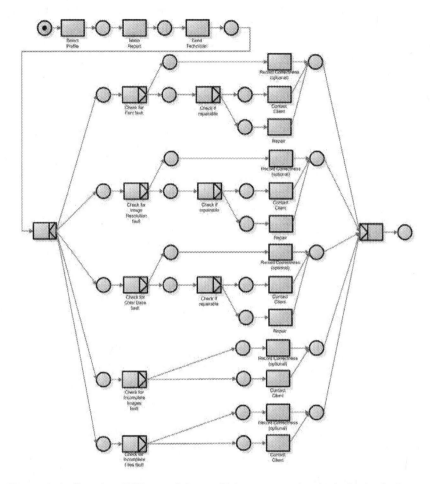

Figure 4–4: Complete WF-net of the preflight process after include the fault treatment subprocess

parallel, (figure 4–5.b) assigning the fault probability to the weight of one of the immediate transitions and its complement to the other one. The resultant model replacing the OR-split blocks is shown in figure 4–6. This net is a Quantitative Dependability WF-net based Model (QDWM).

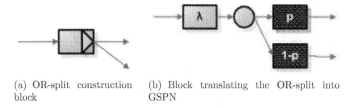

(a) OR-split construction block

(b) Block translating the OR-split into GSPN

Figure 4–5: OR-split translation into GSPN

4.2 The *Ripping* subprocess

Our second case of study is the DP pre-press process of Ripping. Although in this stage the job is supposed to be flawless, in many cases it is not true. Once the job income to the process and it is rippped, the output file is verified to check the correspondence between the intended final print and the print provided by that final ripping output. There are a series of steps that can be done before the ripping process in order to avoid a failure, and after a failure in that ripping process occurs [2] [29]. We are considering in this part of our analysis and for the inclusion into our workflow the most relevant of those correcting steps. Table 4–2 shows those errors present in a document that could cause a failure in the ripping stage. Errors are classified by the kind of failure that they are able to generate.

Table 4–2: Analysis of the possible errors present in ripping

	Domain	Consistency	Consequences
Corrupted copies of fonts	Content/Timing	Inconsist.	Medium
Excessive amount of pages	Timing	Consist.	Medium
Wrong format for image effects	Content/Timing	Inconsist.	Minor
Wrong document generation	Timing	Consist.	Medium

In the same way that we analyze the preflight pre-press process, we first develop an alternative model to measure dependability of the ripping pre-press process.

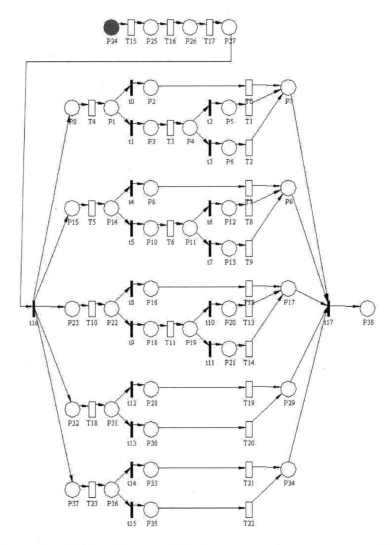

Figure 4–6: Workflow Model of the Preflight pre-press process based on GSPN

Then, we illustrate our methodology to create a workflow model able to measure quantitative dependability attributes applied to the ripping process.

4.2.1 Alternative model to measure dependability applied to the Ripping process

In this case of study we assume that the pre-press process of ripping may fail with a certain probability, that event is called in the fault tree event E. Once the process fails, we also assume that it has four error sources that could lead to a failure of the system, i.e., the ripping pre-press process do not generate correctly the bitmap file for rendering on an output device. Figure 4–7 shows the fault tree representation of the ripping subprocess. The couple of events A, B, C, and D represent those errors, present in the job, that could lead to a failure. We assume that all of them may be fixable with a certain probability. A represents errors related with corrupted copies of fonts, B errors related with excessive amount of pages to be ripped, C errors related with complex image effects not converted to bitmap format, and D errors related with a wrong document generation. The first event of each couple represents the presence of the error in the job and the second event of the couple represents the ability of repair that kind of error. The event F represents the possible case when the existence of every possible error in the job is verified but not founded. In such a case the ripping process is still unsuccessful and it must be labeled as failure.

The analytic equation that describes the probability of failure of the ripping pre-press process, analyzing the fault tree in figure 4–7, was obtained. Thus, the expression of the probability of failure of the ripping stage is the equation (4.3). We used the corresponding letter instead of use the events multiplication (for instance: $A \Rightarrow A_1 \cdot A_2$)

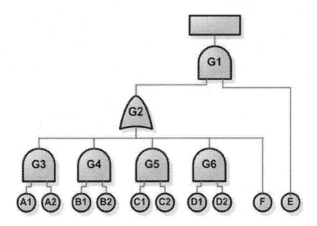

Figure 4–7: Fault Tree for the Ripping pre-press process

$$P_F = [(F + A + B + C + D - FA - BC - BD - CD + BCD)-$$
$$-(F + A - FA) \cdot (B + C + D - BC - BD - CD + BCD)] \cdot E$$

(4.3)

The equation for the reliability of the ripping stage is: $R = 1 - P_F$. Replacing into this equation, we have:

$$R = 1 - [(F + A + B + C + D - FA - BC - BD - CD + BCD)-$$
$$-(F + A - FA) \cdot (B + C + D - BC - BD - CD + BCD)] \cdot E$$

(4.4)

4.2.2 Develop a WF-net including fault parameters for Ripping

To analyze quantitatively dependability attributes in the ripping WF-net, it is necessary to add the identified possible sources of error of the system into the model generated. We include those errors into the model of figure 3–5 by replacing the task *check and repair errors in the job* by four parallel sub-processes, using an AND-split/AND-join construction. Each of them check the presence of a kind of one of the most relevant errors that a job could have. The resulting net is shown in figure 4–8

40

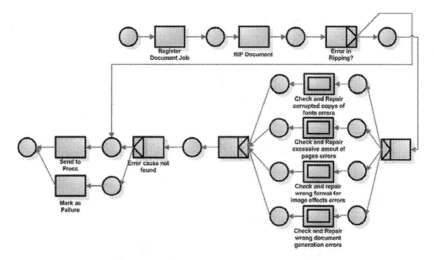

Figure 4–8: Replacing the subprocess "check and repair errors in the job" in the ripping stage WF-net (figure 3–5)

Using the same methodology of replacing the subprocesses of figure 4–8 by the subprocess shown in figure 4–3 we obtain the WF-net shown in figure 4–9. Once again we test soundness in this WF-net using WoPeD and Woflan. In order to convert this net into a GSPN we replace the OR-split blocks (figure 4–5.a) by a combination of transitions. The outcome is governed by a fault probability . The OR-split block (figure 4–5.a) is replaced by an exponentially transition simulating the action followed by two immediate transitions disposed in parallel, (figure 4–5.b) assigning the fault probability to the weight of one of the immediate transitions and its complement to the other one. The resultant model replacing the OR-split blocks is shown in figure 4–10. This net is a Quantitative Dependability WF-net based Model (QDWM) for the DP pre-press process of Ripping.

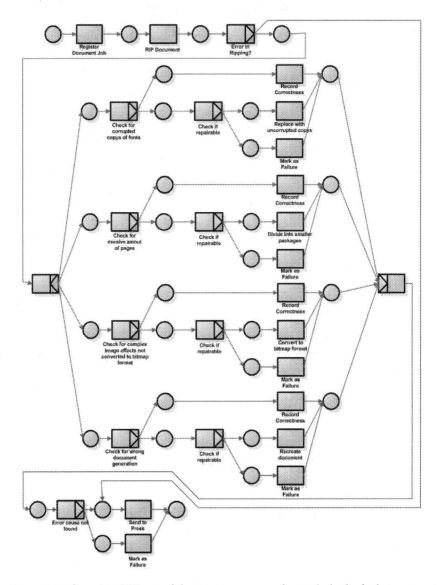

Figure 4–9: Complete WF-net of the ripping process after include the fault treatment subprocess

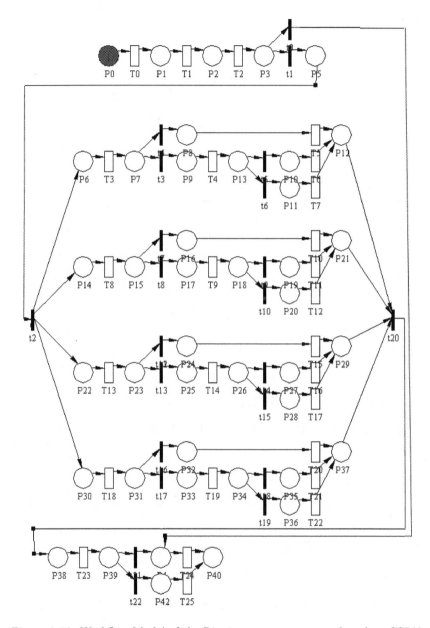

Figure 4–10: Workflow Model of the Ripping pre-press process based on GSPN

CHAPTER 5
COMPARATIVE MEASURES, RESULTS AND ANALYSIS

We have shown two methodologies to analyze dependability on a system so far. The first one is a fault tree model, which is a combinatorial model, that involves the faults, errors and the final system conditions to reach a failure. This combinatorial model gives the opportunity of analyze the quantitative attributes but it does not give the opportunity to analyze either the performance or the structure of the system.

The second one, our proposed methodology, incorporates the combinatorial model of the system based on the faults that this model itself could contain into a workflow model preserving the characteristic routing rules, and behavior of a workflow model, enhancing the model. The resultant net will contain not only fault and failure states, but also normal workflow states, making the model more accurate. Besides structure, behavior, performance, and dependability could be analyzed using the same net.

In this section the results from a QDWM are compared with the results obtained from a Fault Tree model. These results are obtained introducing different vectors of fault and error parameters to each model and comparing the outputs.

To obtain results from the Fault Trees of the Figs. 4–1 and 4–7, vectors above mentioned are introduced in the equations (4.2) and (4.4) respectively. To obtain the quantitative dependability attributes of the GSPN of Figs. 4–6 and 4–10, it is necessary to follow a series of steps. First, it is necessary to generate the reachability graph of the GSPN. From this reachability graph it is possible to deduce the associated Continuous Time Markov Chain (CTMC). Doing the steady-state analysis of the CTMC is obtained the probability of the system of being in any of its states, thus it is achievable to work out the dependability attributes of the system checking the corresponding combination of probabilities of being in certain states that belongs to each attribute. This analysis was done using a software tool named SHARPE [30]. SHARPE is used to model and validate distributed systems using GSPN, among other kind of models. This tool offers a multi-environment graphical interface and provides a specification language and solution methods for performance and reliability modeling.

5.1   Software Tool

We have developed a software tool that consists in two different applications. The first one allows users to create GSPN as well as create, save, and load Petri Nets for Workflow Modeling, i.e., QDWMs. First, the user creates a graphical representation of the QDWM on screen, then each primitive's corresponding attribute is introduced. The application name *task* to a timed transition and *transition* to an immediate transition. The user must specify the name of the primitive (for instance: task1, transition1 or place1), the value of the attribute (number of tokens, time parameter for tasks or weight/probability for transitions), and if the task is a discard task. The application name *discard task* to a task that represents an action of discard or refuse the case treated in the workflow. This task represents an event when the fault or error cannot be repaired or removed from the case becoming into a high probable possibility of failure.

A dummy place is introduced into the net and every task that treats the case of non-repairable fault or error (a discard task) is connected to this place. Subsequently, it is analyzed the probability that this place is empty in steady-state. Thus, we obtain the reliability of the whole process in steady-state. For instance, in figure

4–6, transitions $T2$, $T9$, $T14$, $T20$, and $T22$ would be connected to a dummy place (not depicted).

The QDWM-Creation application shows the net graphically in addition to saving all of the attributes for further processing. The purpose of this tool is to generate a text description of the topology and its attributes (called Netlist) that shall be saved in a file (SHARPE_NETLIST.txt).

The QDWM-Creation application permits the user to interact with a dynamic Graphical User Interface (GUI) on which the Petri Net can be drawn and its data entered. The GUI consists of a window with two sections within it. On the top are eight different buttons each executing a different function. These functions include selecting a primitive to be drawn (Place, Transition, Task, or Arc), selecting a data entry mode, creating a netlist of the net, saving the Petri Net onto file, and loading a previously saved Petri Net. The other section is an area with white background that is left for drawing the Petri Net. In figure 5–1 we see an illustration of the Simulator's GUI.

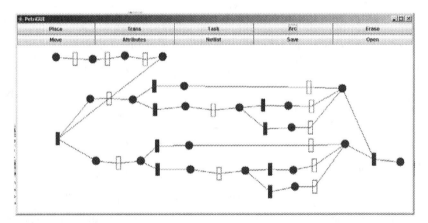

Figure 5–1: Graphical User Interface (GUI) of Simulator application. On the top we see the function buttons, and on the bottom we can see the drawing area

A user creates a Petri Net by selecting primitives from the buttons and clicking on the drawing area. Depending on which primitive is selected at the moment, and by clicking the drawing area the primitive shall appear on screen. After drawing the net, the user can input data into each primitive. This is done by first placing the application in a data entry mode and pressing on the attributes button and right clicking on the primitive. Once this is done, a pop up menu appears with fields for entering name, attribute, and if the primitive is a task, to set it as discarded or not. In Figure 5–2 we see an illustration of the pop up menu. After the Petri Net is created and its data is entered the user can either save the Petri Net or create the Petri Net's netlist on a file. By selecting the Netlist button a file is generated called SHARPE_NETLIST.txt that contains the text description of the topology and attributes of the Petri Net.

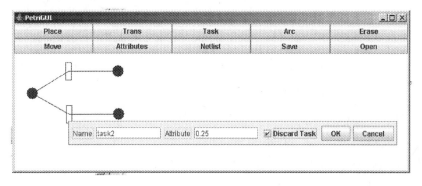

Figure 5–2: Pop up menu for editing primitive attributes. This menu is called when a right click is performed over a primitive and the program is set into data entry mode (Attribute button is clicked)

This file appears in the Project Folder's directory. The Petri Net is can be saved for future use by clicking the Save button another file is generated with the name the user desires. This file contains an internal description, used by this application, of the Petri Net and its data in order to draw and load the net at a later time. To load a Petri Net simply press the load button. A menu shall appear where the user selects which file is to be loaded. Once the file is selected, the Petri Net represented in that file is drawn and its data is loaded into the program's data structures.

The second application of the software is intended to to a batch processing over the QDWM. The file SHARPE_NETLIST.txt is asked to be loaded by the user and it is displayed in the left part of the GUI. The user must select a task or a transition and introduce its name in the text space named "Task or Transition Name". Subsequently the user selects start value, end value and step of the attribute of that primitive first selected. By pressing the button "Perform Calculations" the application feed the analyzer tool SHARPE with the file SHARPE_NETLIST.txt, changing the value of the corresponding primitive. In the center of the GUI are shown the couples of data: Primitive value vs. SHARPE result, for each primitive value. In the right side of the GUI are displayed the outputs of SHARPE for each primitive value. Figure 5–3 shows an screen-shot of the Batch-Processing application. By pressing the button "Graph with Ptplot" the application calls an external application named Ptplot. Ptplot is a software tool developed in the Ptolemy project at the University of California at Berkely and it is a 2D data plotter and histogram tool implemented in Java [31]. Ptplot draws those points into a single graph and connect them and it allows to save the graph or export it to encapsulated postscript format. Fig 5–4 shows the Batch-Processing application calling Ptplot.

5.2   Analysis for the *Preflight* subprocess

For our first case of study, the eight fault parameters on each vector are: Probability of find a fault related with fonts, image resolution, wrong color base, incomplete or corrupted files and, incomplete or missing images. Besides, probability of not repair faults related with fonts, images resolution, and wrong color bases. For each model the fault parameters are the same. As we mentioned before, in the fault tree case, outputs are obtained from the equation (4.2), whereas in the QDWM case, outputs are acquired from the analysis software tool.

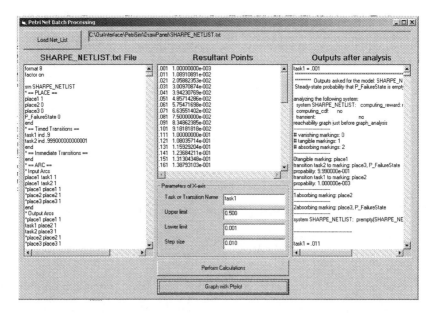

Figure 5–3: Batch-Processing application screen-shot

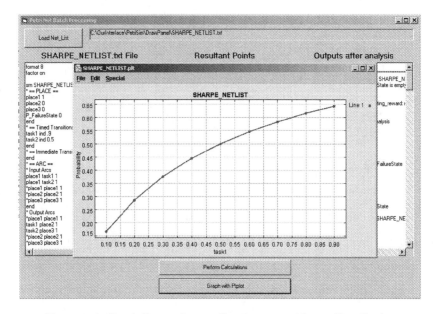

Figure 5–4: Batch-Processing application screen-shot calling Ptplot

Using ten different vectors generated randomly, we can see that the steady-state reliability obtained from the QDWM of the preflight pre-press process, is totally related with the reliability obtained from the Fault Tree model. The mean error between both measures is around $8.0 \cdot 10^{-5}$ and the correlation between them is 0.9999, which is highly close to one. For these reasons, we can tell that a QDWM is able to measure reliability from a process. Due to its WF-net properties were not altered, this QDWM also allows to measure performance attributes of the process and it conserves its main intended function: to be a workflow model.

Concerning to our case of study of preflight, we assume a input vector in order to analyze the reliability of the preflight pre-press process. This vector is as follows: probability of find a fault related with fonts, 0.67; image resolution, 0.39; wrong color base, 0.61; incomplete or corrupted files, 0.05; and incomplete or missing images, 0.05. Besides, probability of not repair faults related with fonts, 0.08; image resolution, 0.05; and wrong color base, 0.03. The resulting reliability is 0.82. The reliability for this pre-press stage is around 0.90 according to local printshops.

5.3 Analysis for the *Ripping* subprocess

For our second case of study, the eight then error events on each vector are: Probability of find corrupted copies of fonts, excessive amount of pages to be ripped, complex image effects not converted to bitmap format, and wrong document generation. Also, probability of cannot replace the file with uncorrupted copies of fonts, cannot divide the job into smaller ripping packages, cannot convert to bitmap format complex image effects before placing them in the application file, and cannot recreate document in a different application. In addition, probability of ripping failure and probability of cannot find the cause of the ripping failure. Once again, for each model the error events are the same. As we mentioned before, in the fault tree case, outputs are obtained from the equation (4.4), whereas in the QDWM case, outputs are acquired from the analysis software tool.

For the case of ripping, we used another group of ten different vectors generated randomly. Again we see that the steady-state reliability obtained from the QDWM model is quite similar with the reliability obtained from the Fault Tree model. The mean error between both measures is also around $8.0 \cdot 10^{-5}$ and their correlation is 0.99999. In this model, the WF-net properties were not altered either, therefore this QDWM also conserves its main intended function: to be a workflow model.

CHAPTER 6
CONCLUSION AND FUTURE WORK

6.1 Concluding remarks

We obtained values of dependability attributes in a quantitative form from the workflow model instead from a combinatorial model adapting some methodologies and strategies for combinatorial model build to the creation of the workflow process definition. This approach will lead to the addition of parameters to the model, in order to make it more versatile. Contrasting the analysis provided in other research projects, our work proposes an analysis of quantitative dependability over a workflow model, and not only an analysis of its structure and performance.

The resulting WF-net from the inclusion of the fault treatment subprocess into the initial workflow model is the QDWM once it is translated into GSPN. Every subprocess used to create these WF-nets is sound, and those nets were tested using specialized software. Accordingly, the QDWM created is completely sound. Therefore, the QDWM allows to analyze the dependability of the system and its methodology of creation preserves its routing structures and its function to measure performance too.

It was shown that structural and performance analysis are important in a workflow model in order to guarantee the best workflow process definition to be implemented in a workflow management system. It is also important to measure dependability attributes and refine the design of that workflow process definition. Dependability attributes have been measured using combinatorial models based on the faults and their probability of occurrence in the system. The idea proposed is to measure quantitatively dependability attributes from the workflow model itself, improving the analysis tools necessary to achieve a good workflow process definition. Applying this new measure concepts to the general Digital Printing pre-press process ensures a better workflow management in this area, because this process is highly based on its trustworthiness. Those measures of dependability are intended to help in the design of a more reliable system.

6.2 Future work

- This work was focused on measuring quantitative dependability attributes. It was achieved by doing a Fault Forecasting analysis. In a future work the Digital Publishing workflow could be studied by using Fault Tolerant analysis (make the system strong enough to detect a fault or an error, and recover from it by itself). One possibility is to apply common techniques such as redundancy and exception handling to the creation of the DP workflow model, so that the business process definition could be enhanced even more. Our proposed workflow does not treat widely the failure situations that may occur. In most cases the proposed solution consist in send back the job to an early stage, which means to start over the process again, or simply to send it back to the client. A fault tolerance workflow would be able to recognize the fault, the error or the failure, and for instance, isolate it. It could be seen as part of go deeper in the study of repairing processes. Exception handling in a workflow could be seen as treat certain cases in a different way. In our proposed workflow, if a job does not have every font embedded, the preflight subprocess will fail. And exception handling could be to mark as successful job in the preflight stage if only a few group of fonts cannot be embedded

- We have created a generic DP workflow model with the most basic and relevant stages and characteristics of the DP business process. However, this workflow could be studied adding some new features of the business process, such as a soft-proofing stage and an accounting stage. The inclusion of new features and stages into the

DP workflow will lead to add new tasks and, therefore, new routing definitions for these tasks. For instance, an accounting stage could calculate job price based on new features settle in the intent stage. If the client decline a preflight service on his job, making lower the printing cost, just a soft-proofing preview could be offered.

- To analyze each subprocess we have use fixed probabilities for all documents. Another possibility is to extend the study to subclassify documents in classes such as books, posters, or magazines, and adapt the probabilities according to the the kind of job. Having those probabilities, the workflow model could be designed to be dynamic, changing the routing paths depending on the case. A workflow management system could be able to manage different workflow engines with different workflow models. These models would depend on a subclassification of the cases treated in the whole system, which lead to characterize the process for each kind of document and gather information about time of completion of task treating only certain kind of documents, probability of fault existence in certain kind of documents, and so on. A fault forecasting analysis would have to be done over each subclass.

APPENDICES

APPENDIX A
SOFTWARE TOOL CODE - SERVICE
MANUAL

A.1   System Requirements

For Windows or Linux, the software requires the use of the following libraries:

- java.awt.*

- java.awt.event.*

- javax.swing.*

- java.util.*

- java.io.*

The application will run on any OS that possesses an up to date Java Virtual Machine (JVM).

To EDIT or RUN application:

Java NetBeans IDE 4.1- This program was written and tested using IDE. Simply open a project with the source code and package of this application. It can be proceeded to edit, build and run the application from this environment. Trying to edit and run this application in a more recent NetBeans IDE (5 and up) has caused problems. In this case a new project should be opened and only the source code files should be imported into the project on the new IDE.

Petri Net Simulator Tool should also run on any Java Interpreter. In that case the class files (x.class) from this application should be run.

APPENDIX B
SOFTWARE TOOL - USER MANUAL

B.1 Getting Started

Petri Net Simulator Tool is a software application written in JAVA that allows the user to create, save, and load Petri Nets for Workflow Modeling. The Petri Net consists of its graphical representation on screen along with its primitive's corresponding attributes. The simulator shows the net graphically in addition to saving all of the attributes for further processing. The purpose of this tool is to generate a text description of the topology and its attributes (called Netlist) that shall be saved in a file (SHARPE_NETLIST.txt) and later on this file is entered into an analyzer tool (SHARPE) for further processing.

B.2 Petri Net Simulator Features

The Simulator application permits the user to interact with a dynamic Graphical User Interface (GUI) on which the Petri Net can be drawn and its data entered. The GUI consists of a window with two sections within it. On the top are eight different buttons that each execute a different function. These functions include selecting a primitive to be drawn (Place, Transition, Task, or Arc), selecting a data entry mode, creating a netlist of the net, saving the Petri Net onto file, and loading a previously saved Petri Net. The other section is an area with white background that is left for drawing the Petri Net. In figure B–1 we see an illustration of the Simulator's GUI.

A user creates a Petri Net by selecting primitives from the buttons and clicking on the drawing area. Depending on which primitive is selected at the moment, and

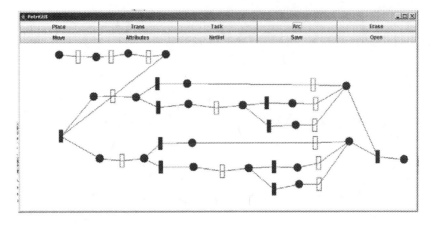

Figure B-1: Graphical User Interface (GUI) of Simulator application. On the top we see the function buttons, and on the bottom we can see the drawing area

by clicking the drawing area the primitive shall appear on screen. After drawing the net, the user can input data into each primitive. This is done by first placing the application in a data entry mode by pressing on the attributes button and right clicking on the primitive. Once this is done, a pop up menu appears with fields for entering name, attribute, and if the primitive is a task, to set it as discarded or not. After the Petri Net is created and its data is entered the user can either save the Petri Net or create the Petri Net's netlist on a file. By selecting the Netlist button a file is generated called SHARPE_NETLIST.txt that contains the text description of the topology and attributes of the Petri Net. This file is what the analyzing tool SHARPE takes as input to perform calculations of dependability for that net. This file appears in the Project Folder's directory. If the Petri Net is to be saved for future use, by clicking the Save button another file is generated with the name the user desires. This file contains an internal description, used by this simulator, of the Petri Net and its data in order to draw and load the net at a later time. To load a Petri Net simply press the load button. A menu shall appear where the user selects which file is to be loaded. Once the file is selected, the Petri Net represented in that file is drawn and its data is loaded into the program's data structures.

B.3 Functions

Following is a description of each button and its function.

- Place Button- By selecting the place button the program is set to draw a Place primitive on the drawing area. After this button is selected, a click on the drawing area will create a place on screen.

- Trans Button- By selecting the transition button the program is set to draw a Transition primitive on the drawing area. After this button is selected, a click on the drawing area will create a transition on screen.

- Task Button- By selecting the task button the program is set to draw a Task primitive on the drawing area. After this button is selected, a click on the drawing area will create a task on screen.

- Arc Button- By selecting the Arc button the program is set to draw an Arc primitive on the drawing area. After this button is selected, a press of the mouse on the drawing area will begin to create a line on screen. The mouse must be thereon dragged and the arc will still be rendered. Once the mouse reaches the arc's endpoint the mouse is released and the arc is finalized with its starting point and ending point.

- Erase Button- By selecting the Erase button the program is set into deletion mode. After this button is selected a left click on top of a Petri Net primitive shall not only erase the primitive, but also erase any arcs that are connected to it.

- Move Button- By selecting the Move button the program is set into a selection mode. After this button is pressed the user can select, drag, and drop any primitive to a new position. Any arcs that are attached to the primitive being moved remain connected to it, and automatically reposition themselves to the primitive's new position.

- Attributes Button- By selecting the Attributes button the program is set into a data entry mode. After this button is selected a right click on top of a Petri Net primitive will call a pop up menu which contains fields for entering name, attribute, and if the primitive is a task, to set it as discarded or not by means of a checkbox. Once the data is entered in the pop up menu, pressing the OK button of the pop up menu saves the primitive's data and returns from the menu. If the Cancel button is selected the program returns from the pop up menu without saving any data. In Figure B–2 we see an illustration of the pop up menu.

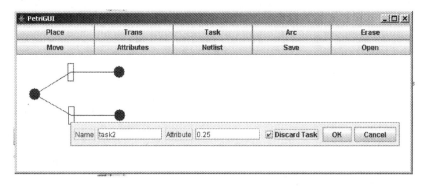

Figure B–2: Pop up menu for editing primitive attributes. This menu is called when a right click is performed over a primitive and the program is set into data entry mode (Attribute button is clicked)

- Netlist Button- By selecting the netlist button a file is generated called SHARPE_NETLIST.txt and is saved in the project folder's directory. This file is written in a special format that SHARPE understands. This file contains a text description of the Petri Net's topology and data. If there is already a file called SHARPE_NETLIST.txt in that path it shall be overwritten.
- Save Button- By selecting the save button a menu appears in which the user selects the path and file name where to save the Petri Net. Once a file is selected, the program generates in this file a text description of the Petri Net's topology and its data. This is done to load a previously saved Petri Net at a later time.

- Open Button- By selecting the open button a menu appears in which the user selects the path and file of the Petri Net that is to be loaded. Once the appropriate file is selected, the program reads and interprets its contents, proceeds to draw the whole net and load its attributes into the program's data structures.

REFERENCE LIST

[1] N. Santiago, F. Vega-Riveros, W. Rivera, M. Rodriguez-Martinez, T. Avellanet, G. Chaparro-Baquero, W. Lozano, A. Pereira, and H. Santos-Villalobos. Towards development of concepts and algorithms to enable automated digital publishing workflows. Technical report, University of Puerto Rico, Mayaguez Campus, May 2005.

[2] M. L. Kepler. *The Handbook of Digital Publishing - Vol. II.* Prentice-Hall, 2001.

[3] D. Monkerud. Realizing the promise of workflow automation. *Digital Publishing Solutions Magazine Online*, May 2004.

[4] R. Zurawski and M. Zhou. Petri nets and industrial applications - a tutorial. *IEEE Trans. on Industrial Electronics*, 41(6):567–583, 1994.

[5] W. M. P. van der Aalst and K. M. van Hee. *Workflow Management: Models, Methods, ans Systems.* The MIT Press, Cambridge, Massachusetts, 2002.

[6] A. Avizienis, J.C. Laprie, B. Randell, and C. Landwehr. Basic concepts and taxonomy of dependable and secure computing. *IEEE Transactions on Dependable and Secure Computing*, 01(1):11–33, January-March 2004.

[7] F. Bause and P. Kritzinger. *Stochastic Petri Nets, an introduction to the theory.* Friedr. Vieweg & Son, 2002.

[8] W. M. P. van der Aalst. The Application of Petri Nets to Workflow Management. *The Journal of Circuits, Systems and Computers*, 8(1):21–66, 1998.

[9] B. Mikolajczak and D.L. Byrne. Workflow modeling and diagnosis with petri nets - a case study of a manufacturing process. In *Systems, Man and Cybernetics, 2002 IEEE International Conference on*, page 6 pp. vol.5, October 2002.

[10] J. Li, Y. Fan, and M. Zhou. Performance modeling and analysis of workflow. *Systems, Man and Cybernetics, Part A, IEEE Transactions on*, 34(2):229– 242, March 2004.

[11] W. M. P. van der Aalst and B. F. van Dongen. Discovering workflow performance models from timed logs. In *Engineering and Deployment of Cooperative Information Systems, First International Conference (EDCIS 2002) Beijing, China, September 17-20, 2002 / Y. Han, S. Tai, D. Wikarski (Eds.)*, pages 1–45pp. Springer Verlag, LNCS 2480, September 2002.

[12] S. Ling and H. Schmidt. Time petri nets for workflow modelling and analysis. In *Systems, Man, and Cybernetics, 2000 IEEE International Conference on*, pages 3039–3044 vol.4, 2000.

[13] G. Alonso, C. Hagen, D. Agrawal, A. E. Abbadi, and C. Mohan. Enhancing the fault tolerance of workflow management systems. *IEEE Concurrency*, 8(3):74–81, 2000.

[14] A. Avizienis, J. C. Laprie, and B. Randell. Fundamental concepts of dependability. In *3rd Information Survivability Workshop (ISW'2000), Boston (USA)*, pages 7–12, October 2000.

[15] A. Johnson Jr. and M. Malek. Survey of software tools for evaluating reliability, availability, and serviceability. In ACM Press, editor, *ACM Computing Surveys (CSUR)*, volume 20, pages 227–269, December 1988.

[16] N. B. Fuqua. The applicability of markov analysis methods to reliability, maintainability, and safety. *Selected Topics in Assurance Related Technologies START*, 10(2), 2003.

[17] M. Nicholson and J. McDermid. Analysis of dependable computer systems. Technical report, University of York, November 1994.

[18] D.I. Heimann, N. Mittal, and K.S. Trivedi. Dependability modeling for computer systems. In *Reliability and Maintainability Symposium, 1991. Proceedings., Annual*, pages 120–128, January 1991.

[19] N. Lopez-Benitez. Dependability analysis of distributed computing systems using stochastic petri nets. In *Reliable Distributed Systems, 1992. Proceedings., 11th Symposium on*, pages 85–92, October 1992.

[20] J. K. Muppala and C. Lin. Dependability analysis of large-scale distributed systems using stochastic petri nets. In *Systems, Man, and Cybernetics, 1996., IEEE International Conference on*, pages 3033 – 3038, October 1996.

[21] J. C. Laprie. Dependability of computer systems: concepts, limits, improvements. In *Software Reliability Engineering, 1995. Proceedings., Sixth International Symposium on*, pages 2–11, October 1995.

[22] A. Bondavalli, I. Majzik, and I. Mura. Automatic dependability analysis for supporting design decisions in uml. In *High-Assurance Systems Engineering, 1999. Proceedings. 4th IEEE International Symposium on*, pages 64–71, 1999.

[23] D. Tang, M. Hecht, J. Agron, A. Miller, and H. Hecht. Engineering oriented dependability evaluation: Meadep and its applications. In *Fault-Tolerant Systems, 1997. Proceedings., Pacific Rim International Symposium on*, pages 85–90, December 1997.

[24] S. Asgari, V. Basili, P. Cost, P. Donzelli, L. Hochstein, M. Lindvall, I. Rus, F. Shull, R. Tvedt, and M. Zelkowitz. Empirical-based estimation of the effect on software dependability of a technique for architecture conformance verification. In *Workshop on Architecting Dependable Systems, ICSE, Edinburgh, Scotland*, May 2004.

[25] R. Rao, G. Swaminathan, B.W. Johnson, and J. H. Aylor. Synthesis of reliability models from behavioral-performance models. In *Proceedings of the 1994 Reliability and Maintainability Symposium (RAMS)*, pages 292–297, January

1994.

[26] R. H. Klenke, M. Meyassed, J. H. Aylor, B. W. Johnson, R. Rao, and A. Ghosh. An integrated design environment for performance and dependability analysis. In *DAC '97: Proceedings of the 34th annual conference on Design automation*, pages 184–189, New York, NY, USA, 1997. ACM Press.

[27] University of Cooperative Education (Berufsakademie) Karlsruhe. Woped ("workflow petrinet designer"). http://www.woped.org, Accessed: March-2006.

[28] H.M.W. Verbeek, T. Basten, and W.M.P. van der Aalst. Diagnosing workflow processes using woflan. In *The Computer Journal. British Computer Society*, volume 44, pages 246–279, 2001.

[29] J. Howard B. It will not rip, rip errors and fixes. http://desktoppub.about.com/cs/rip/a/rip_2.htm, May 2006.

[30] C. Hirel, R. Sahner, X. Zang, and K. Trivedi. Reliability and performability modeling using sharpe 2000. In *Proceedings of the 11th International Conference on Computer Performance Evaluation: Modelling Techniques and Tools*, volume 1786, pages 345–349.

[31] E. A. Lee. Overview of the ptolemy project. Technical Memorandum UCB/ERL M03/25, University of California, Berkeley, July 2003.

BIOGRAPHICAL SKETCH

Gustavo Adolfo Chaparro-Baquero was born in January 22th of 1980 in Bogotá, Colombia. Gustavo studied his bachelor in the Pontificia Universidad Javeriana at Bogotá, obtaining his title as Electronic Engineer in September of 2002. His grading project was an "Educational Software for Children with Cerebral Paralysis". His work was meritorious to win the National Award for the Technology Innovation Granted by E.T.B. and The National University of Colombia, in the category of Innovation on Information Technologies pursuing the social improvement in Colombia in 2003. Gustavo received his title as Master of Science in Computer Engineering from the University of Puerto Rico, Mayaguez Campus in 2006. A paper from his Master's Thesis was published in the 2nd IEEE International Symposium on Dependable, Autonomic and Secure Computing (DASC'06), Indiana University, Purdue University, Indianapolis, USA September 29 - October 1, 2006.

PETRI NET WORKFLOW MODELING FOR DIGITAL PUBLISHING MEASURING QUANTITATIVE DEPENDABILITY ATTRIBUTES

Gustavo Adolfo Chaparro-Baquero
(787) 484-4953
Department of Electrical and Computer Engineering
Chair: Nayda G. Santiago, Ph.D
Degree: Master of Science
Graduation Date: July 12th 2006

This work describes the concept of workflow modeling using Generalized Stochastic Petri Nets (GSPN) for the Digital Publishing business process and how the attributes of dependability are measured in a quantitative form. In our novel approach, these are measured from the workflow model itself, improving the analysis of a workflow model. Applying these measure concepts to the general Digital Publishing pre-press process provides a better workflow management in this area than the actual procedure, since this process is based on its trustworthiness. Once the methodology for workflow modeling measuring quantitative dependability attributes is introduced, the results for a case study on the preflight stage and for the ripping stage of the Digital Publishing workflow are presented.

www.ingramcontent.com/pod-product-compliance
Lightning Source LLC
LaVergne TN
LVHW080103070326
832902LV00014B/2400